T5-ARF-754

BUT EVEN SO

BY KENNETH PATCHEN

Aflame and Afun of Walking Faces
An Astonished Eye Looks Out Of The Air
A Surprise For The Bagpipe-Player
Because It Is
Before The Brave
But Even So
Cloth Of The Tempest
Doubleheader
Fables & Other Little Tales
First Will & Testament
Glory Never Guesses
Hurrah For Anything
Memoirs Of A Shy Pornographer
Orchards, Thrones & Caravans
Panels For The Walls Of Heaven
Pictures Of Life And Of Death
Poemscapes
Poems Of Humor & Protest
Red Wine & Yellow Hair
See You In The Morning
Selected Poems
Sleepers Awake
The Collected Poems of Kenneth Patchen
The Dark Kingdom
The Famous Boating Party
The Journal Of Albion Moonlight
The Love Poems of Kenneth Patchen
The Teeth Of The Lion
They Keep Riding Down All The Time
To Say If You Love Someone
When We Were Here Together

But Even So

KENNETH PATCHEN

A NEW DIRECTIONS BOOK

N
811 6537
P2 .P27
B8

Copyright © 1968 by Kenneth Patchen

Library of Congress Catalog Card Number: 68-9474

All rights reserved. Except for brief passages quoted in a newspaper, magazine, radio, or television review, no part of this book may be reproduced in any form or by any means, electronic or mechanical, including photocopying and recording, or by any information storage and retrieval system, without permission in writing from the Publisher.

Manufactured in the United States of America.

Published simultaneously in Canada by McClelland & Stewart, Limited.

New Directions Books are published for James Laughlin
by New Directions Publishing Corporation,
333 Sixth Avenue, New York 10014.

THIRD PRINTING

Withdrawn from
Cleveland Inst. of Art Library

FOR MIRIAM

73-241

BUT EVEN SO

But Even So

They Don't Seem To Understand That Unless Someone Does Nothing Soon The Sky'll Sure Not Be All That's Up

Patchen

But Even So

go LOVING

Patchen

& THEM-WITH
ALL THEY ARE
OR EVER WERE-
YOU'LL OVERTHROW

But Even So

This room, this battlefield

But

Even

So

Come now, my child
if we were planning
to harm you, do you think
we'd be lurking here
beside the path
in the very dark-
est part of
the forest?
Patchen

But Even So

But Even So

WHAT WALKING
LOOKS
LIKE WHEN
IT'S NOT
And As
It lic kerously
crosses a newly milled pond
to visit Alexander the Grouse
who is
quite un-
querulously
a spot of
swan-
ing
enjoying
Patchen

But Even So

The Hands of the Air

applaud the Wonders of God Light & Life unending

Patchen

But Even So

Ah, Come This Time

Next Never

Things'll Be Fine

Patchen

But Even So

QUIET

WE MUST NOT DISTURB

THE

EVENING-BEING

DEVICE

Petchum

But Even So

CAN'T RECALL ME ONE REASON
FOR GIVIN' A DAM,
WHERE "WHERE WENT"

What a sweet buncha pigs.
Bloody footed hypocrites
Fouling up everything in sight
While you blather about saving
The world—with all the sincerity
And good will   of a snake

Shaking hands with a rabbit
But don't get me wrong
As bastards go, you're long overdue
The hell is, you'll be taking the
Whole big beautiful world with you

Petcham

But Even So

KING

JIZ

Patchen

WONDERS WHICH OF HIM ISN'T HIS

But Even So

But Even So

But Even So

But Even So

Yeah, Jack, it's a dark deck
& a liar's deal
THE CENTURY OF
THE COCKROACH
the handwriting on the wall, eh!
well look at them look at what's
your damned wall!
look at them
roll the credits!
the crapbugs
are adequate
everywhere!
those 30 lousy
pieces of silver
— that was the bribe
— the world!
and even
the stinking, slimy cheaters xxx
Patchen
this is his home base
190,000,000 to one
you open that door
in'll walk some
kind of rubble-talkin'
flag-wrapped whore
Come on you bastards
let them see your faces!
Say something

But Even So

Caring is the only daring oh you know it

Patchen

But Even So

THE DAY HARLEQUIN TO OF

A CATERPILLAR

maybe so it will swivel soon
And the behind also of my mother
Pung! zung! until he is in fronts of
All those others; down the barber's steps
Into Mr Klerp's office — of whom less
A nod. Oh futz and butz never bring
No beer to baby — Bamboozle Nobones
Is one hell of a moniker! I wonder
If anybody's living in that ugly man up there

Patchen

But Even So

IT IS OUTSIDE US
AS WE ARE WITHIN IT

Ecstasy! — "Thinking" is always
outside in the light — THE LIGHT
IS ALIVE! At its touch the "mind" dances!

The Flowing Animal of Light Is All.

The Sleeping-Awakeness — Joy Joying:
The Being-Present Self — But words make mistakes
when the light moves away THE LIGHT IS THE

But Even So

But Even So

Piggykstew

WORLD CHAMP TICKET SELLER

Nary no bus, jitney, streetcar, buggy, aireo-plane — no puffy l'il ole train — not even a ox cart — assumin' that I read you right, chum

— in the whole of your quaint and certainly most unsightly country, that is. Well, then, boy, you've just sold me an upper berth piggy-back to your nearest submarine base

Patchen

But Even So

At one time
the grass was
thought to be
A people some
what like the
— who were
we? Oof. that
ardent necromancy,
the silence a piece of
flesh to praise — un-
issuing epilogue
to the unsaid — upon
a blackened hill.
I was willing for
a garden, oh but
not to have
the blood
of the whole
world, dripping
down upon it
forever

As he
whin
darkes
kom
sa red
Th' old
Lass a
Bread
then rises
f'rae
th' new
Bride's
pure
Bed

But Even So

But Even So

And Mr Eggleg said

You wait there until I get back hear!

and he propped himself up near the telephone and he hurries off; exactly one year later the phone rings – so he unprops and says Hell-lo Mr Eggleg here

I know who I am you fool! what's the address? the building's been torn down – hello! am I still there? (fact is, he is indeed; for Mr Eggleg repropped comfortably, and never was bothered by himself again

MY NAME IS DOBBLE AND I LIVE IN A USED MILK BOTTLE
What's with you, friend?
Patchen

But Even So

Sure, Leroy,
There Is
A
Kindness
Of
Willing
Cruelty
& A Cruelty
Of Won'ting
Still—no dog ever
Doubt his dogness
The maple tree
Never mistake
Himself for
Any other cuss
But a man—
Who is us?

Patchen

But Even So

But
Even
So

But Even So

Mr Plickspoon of Darby & the Doon adair Fly-swatter-Swallower's Daughter's Yoddling-Wooden-Leg Well! iffin now, say, the desire's on you to open a window, why this here's one method guaranteed positively to give the quickest and most complete satisfaction every single time out

Potchen

But Even So

But Even So

But Even So

The Rain Never gets Wet

W. W.

Yet We Patchen Rit Wocket rushes away
And leaves that little ho   rse
Wishing it had   a
Nice mother apple to nu   zzle
OR at least   some
Old train dandruff to sniff

But Even So

As bitter &
far as a tiger's
frown O "Rulers"
of the World

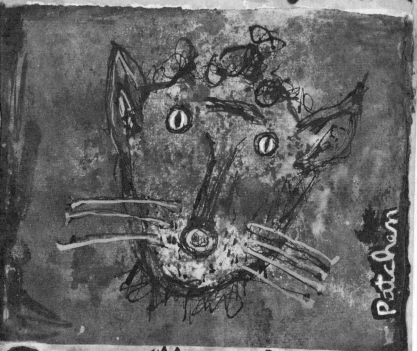

Patchen

So my "blessings" unto you

But Even So

His suchamuch? Name me one other two-legged bug. So he does think the world owes him a dying what's another stupid little mistake? There's lots of room. Besides, he hasn't been here long enough to know his aspect from his ebullience.

Patchen

what he calls "How could he world without it"! — See what I mean?

Take now "speech" run his

But Even So

A Mountain's Knees Seldom Sneeze

Unless he's allergic to the whisdust of some of the more ferocious pine trees

Patchen

But Even So

O O Fountain
Me A
Burning
Bell
O For None
Waken To
The Light
& Cold
Falls
This
Eternity of light

Patchen

But Even So

But Even So

Oh, oh... "Bullet Eye" Braxton,
The Tree-Looker,
has been
by again—
third time today
that oak's
been shriveled
down to the size
of a lil bitty
tobaccah-
chawin'
paw-paw-
bush.
Wonder
-what
Oedipus
would've
made
of this?

Patchen

But Even So

It's outside The Thing
That Dances
not in "our minds"

Dreams "come" in
The Fun of Christ
Wholeness &
Freedom!
The universe is
like a gigantic
bird
hopping
up & down
in such joy
that from time
to time it
just topples over

Patchen

But Even So

But Even So

And Pocahontas
she Done
Cry-
"Bye, son!
Even though you is
a Sniff,
you should'nta
dropped in
on no river
'less'n you water-proofed it first

But Even So

But Even So

How Do You Mourn These Dead?

Who Are Still Alive

Forgive
our
little
children
& all
other
guiltless
helpless
creatures

For we
know
what we do.
Any evil
is good,
provided
it is
OUR
evil!

Under These Christ-Masks, You Prepare

The Murder of the World

In the name of Mankind, of Truth, of Sanity, of Life Itself, I denounce you! as I denounce your "enemies"!

Down with you! Down with all of you! Madmen! Liars! Murderers!

Patchen

But Even So

WE
Deserve Us

While the white lion sat smiling at
his merciless loom, see,
there's this
crownation
of kids
just sitting
down to their
lantern soup,
when this here
old grandma lady
up and squirts them
with her flame-
thrower

Patchen

But Even So

But Even So

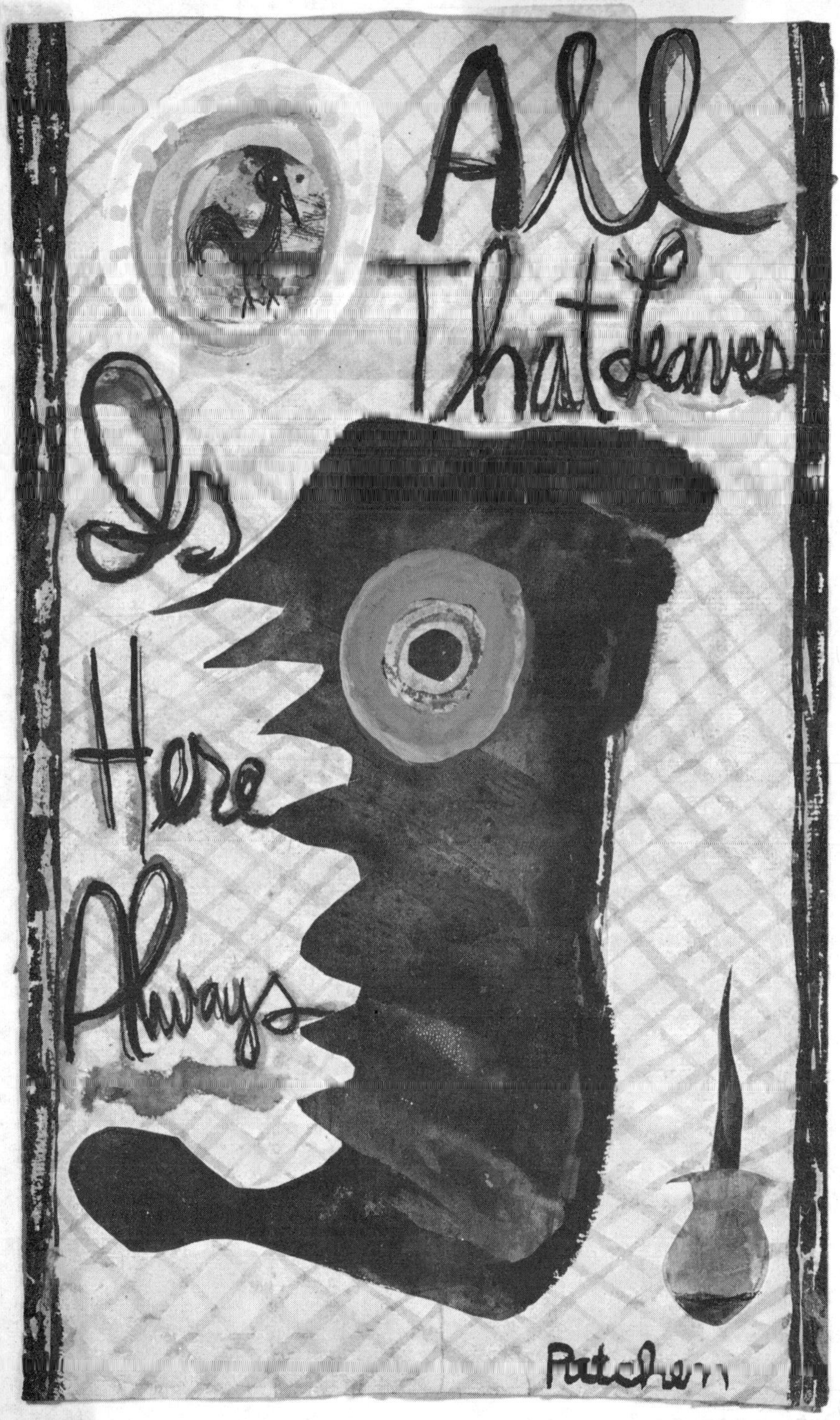
All
That Leaves
Is
Here
Always
Patchen

DATE DUE